The Unconscious Action of the Brain

William B. Carpenter

The Unconscious Action of the Brain

LM Publishers

What goes on in the depths of our own Minds?

I shall begin to reply with an anecdote that was related to me after a lecture which I gave upon this subject about five years ago, at the Royal Institution, in London. As I was coming out from the lecture-room, a gentleman stopped me and said, "A circumstance occurred recently in the north of England, which I think will interest you, from its affording an exact illustration of the doctrine which you have been setting forth to-night." The illustration was so apposite, and leads us so directly into the very heart of the inquiry, that I shall make it, as it were, the text for the commencement of this evening's lecture. The manager of a bank in a certain large town in Yorkshire could not find a key which gave access to all the safes and desks in the bank. This key was a duplicate key, and ought to have been found in a place accessible only to himself and to the assistant-manager. The assistant-manager was absent on a holiday in Wales, and the manager's first impression was, that the key had probably been taken away by this assistant in mistake. He wrote to him, and learned to his own great surprise and distress that he had not got the key, and knew nothing of it. Of course, the idea that the key, which gave access to every valuable in the bank, was in the hands of any wrong person, having been taken with a felonious intention, was to him most distressing. He made search everywhere, thought of every place in which the key might possibly be, and could not find it. The assistant-manager was recalled, both he and every person in the bank were questioned, but no one could give any idea of where the key could be. Of course, although no robbery had taken place up this point, there was the apprehension that a robbery might be committed after the storm, so to speak, had blown over, when a better opportunity would be afforded by the absence of the same degree of watchfulness. A first-class detective was then brought down from London, and this man had every opportunity given him of making inquiries; every person in the bank was brought up before him; he applied all those means of investigation which a very able man of this class knows how to employ; and at last he came to the manager and said: "I am perfectly satisfied that no one in the bank knows anything about this lost key. You may rest assured that you have put it somewhere yourself, and you have been worrying yourself so much about it that you have forgotten where you put it away. As long as you worry yourself in this manner, you will not remember it; but go to bed to-night with the assurance that it will be all right; get a good night's sleep; and in the morning I think it is very likely you will remember where you have put the key." This turned out exactly as it was predicted. The key was found the next morning in some extraordinarily secure place which the manager had not previously thought of, but in which he then felt sure he must have put it himself.

Now, then, ladies and gentlemen, this you may say is merely a remarkable case of that which we all of us are continually experiencing; and so I say it is. Who is there among you who has not had occasion some time or other to try to recall something to his (or her) mind which he has not been able to bring to it? He has seen some one in the street, for instance, whose face he recognizes, and says, "I ought to know that person;" and thinks who it can be, going over (it may be) his whole list of friends and acquaintances in his mind, without being able to recall who it is; and yet, some hours afterward, or it may be the next day, it flashes into his mind who this unknown person is. Or you may want to remember some particular and recent event; or it may be, as I have heard classical scholars say, to recall the source of a classical quotation.

They "cudgel their brains," to use a common expression, and are unsuccessful; they give their minds to something entirely different; and some hours afterward, when their thoughts are far away from the subject on which they had been concentrating them with the idea of recovering this lost clew, the thing flashes into the mind. Now, this is so common an occurrence, that we pass it by without taking particular note of it; and yet I believe that the inquiry into the real nature of this occurrence may lead us to understand something of the inner mechanism of our own minds which we shall find to be very useful to us.

There is another point, however, arising out of the story which I have just told you, upon which again I would fix your attention: Why and how did the detective arrive at this assurance from the result of his inquiries? It was a matter of judgment based upon long practice and experience, which had given him that kind of insight into the characters, dispositions, and nature of the persons who were brought before him, which only those who have got this faculty as an original gift, or have acquired it by very long experience, can possess with anything like that degree of assurance which he was able to entertain. I believe that this particular power of the detective is, so to speak, an exaltation in a particular direction of what we call "common-sense." We are continually bringing to the test of this common-sense a great number of matters which we cannot decide by reason; a number of matters as to which, if we were to begin to argue, there may be so much to be said on both sides, that we may be unable to come to a conclusion. And yet, with regard to a great many of these subjects some of which I shall have to discuss in my next lecture we consider that common-sense gives us a much better result than any elaborate discussion. Now, I will give you an illustration of this which you will all readily comprehend: Why do we believe in an external world? Why do I believe that I have at present before me many hundreds of intelligent auditors, looking up and listening to every word that I say? Why do you believe that you are hearing me lecture? You will say at once that your common-sense tells you. I see you; you see and hear me; and I know that I am addressing you. But if once this subject is logically discussed, if once we go into it on the basis of a pure reasoning process, it is found really impossible to construct such a proof as shall satisfy every logician.

As far as my knowledge extends, every logician is able to pick a hole in every other logician's proof. Now, here we have, then, a case obvious to you all, in which common-sense decides for us without any doubt or hesitation at all. And I venture to use an expression upon this point which has been quoted with approval by one of the best logicians and metaphysicians of our time, Archbishop Manning, who cited the words that I have used, and entirely concurred in them, namely, that "in regard to the existence of the external world the common-sense decision of mankind is practically worth more than all the arguments of all the logicians who have discussed the basis of our belief in it." And so, again, with regard to another point which more nearly touches our subject to-night—the fact that we have a Will which dominates over our actions; that we are not merely the slaves of automatic impulse which some philosophers would make us—"the decision of mankind" (as Archbishop Manning, applying my words, has most truly said), "derived from consciousness of the existence of our living self or personality, whereby we think, will, or act, is practically worth more than all the arguments of all the logicians who have discussed the basis of our belief in it."

Now, then, my two points are these: What is the nature of this process which evolves, as it were, this result unconsciously to ourselves, when we have been either asleep, as in the case of the banker, or, as in the other familiar case I have cited, when we have been giving our minds to some other train of thought in the interval? What is it that brings up spontaneously to our consciousness a fact which we endeavored to recall with all the force of our will, and yet could not succeed?

And then again: What is the nature of this common-sense, to which we defer so implicitly and immediately in all the ordinary judgments of our lives?

Now, in order that we may have a really scientific conception of the doctrine I would present to you, I must take you into an inquiry with regard to some of the simpler functions of our bodies, from which we shall rise to the simpler actions of our minds. You all know that the Brain, using the term in its general sense, is the organ of our Mind. That everyone will admit. We shall not go into any of the disputed questions as to the relations of Mind and Matter; for the fact is, that these are now coming to take quite a new aspect, from Physical philosophers dwelling so much more upon Force than they do upon Matter, and on the relations of Mind and Force, which every one is coming to recognize. Thus, when we speak of nerve-force and mind as having a most intimate relation, no one is found to dispute it; whereas, when we talk about Brain and Mind having this intimate relation, and Mind being the function of the brain, there are a great many who will rise up against us and charge us with materialism, and atheism, and all the other deadly sins of that kind. I merely speak of the relation of the brain to the mind, as the instrument through which the mind operates and expresses itself.

We all know that it is in virtue of the impressions carried to the brain, through the nerves proceeding from the different sensory organs in various parts of the body, that we become conscious of what is taking place around us. And, again, that it is through the nerves proceeding from the brain that we are able to execute those movements which the Will prompts and dictates, or which arise from the play of the Emotions. But I have first to speak of a set of lower centres, those which the Will can to a certain extent control, but which are not in such immediate relation to it as is the brain. You all know that there passes down our backbone a cord which is commonly called the "Spinal Marrow." Now, this spinal marrow gives off a pair of nerves at every division of the backbone; and these nerves are double in function—one set of fibres conveying impressions from the surface to the spinal cord, the other motor impulses from the spinal cord to the muscles. Now, it used to be considered that this Spinal Cord (I use the term spinal cord, which is the same as spinal marrow, because it is just as intelligible and more correct) was a mere bundle of nerves proceeding from the brain; but we have long known that this is not the case, that the spinal cord is really a nervous centre in itself, and that if there were no brain at all the spinal cord would still do a great deal. For example, there have been infants born without a brain, yet these infants have breathed, have cried, have sucked, and this in virtue of the separate existence and the independent action of this spinal cord.

Let us analyze one or two of these actions. We will take the act of Sucking as the best example, because experiments have been made upon young puppies, by taking out the brain, and then trying whether they would suck; and it was found that putting between the lips the finger moistened with milk or with sugar and water, produced a distinct act of suction, just as when an infant is nursed. Now, how is this? It is what we call a "reflex action." I shall have a good deal to say of reflex action higher up in the nervous system, and therefore I must explain precisely what we mean by that term. It is just this: There is a certain part of the spinal cord, at the top of the neck, which is what we call a ganglion, that is, a centre of nervous power: in fact, the whole of the spinal cord is a series of such ganglia; but this ganglion at the top of the neck is the one which is the centre of the actions which are concerned in the act of sucking. Now, this act of sucking is rather a complicated one, it involves the action of a great many muscles put into conjoint and harmonious contraction. We will say, then, that here is a nervous centre. [Dr. Carpenter made a sketch upon the black-board.] These are nerves coming to it, branches from the lips; and these another set going to the muscles concerned in the movement of sucking from it. Thus, by the conveyance to the ganglionic centre of the impression made on the lips, a complicated action is excited, requiring the combination of a number of separate muscular movements. We will take another example—the act of Coughing.

You feel a tickling in your throat, and you feel an impulse to cough which you cannot resist; and this may take place not only when you are awake and feel the impulse, but when you are asleep and do not feel it. You will often find persons coughing violently in sleep, without waking or showing any sign of consciousness. Here, again, the stimulus, as we call it, produced by some irritation in the throat, gives rise to a change in the nerves going toward the ganglionic centre, which produces the excitement of an action in that centre that issues the mandate, so to speak, through the motor nerves to the muscles concerned in coughing, which actions have to be united in a very remarkable manner, which I cannot stop to analyze; but the whole action of coughing has for its effect the driving out a violent blast of air, which tends to expel the offending substance. Thus, when anything "goes the wrong way," as we term it—a crumb of bread, or a drop of water finding its way into the windpipe, then this sudden and violent blast of air tends to expel it.

Now, these are examples of what we call "reflex action;" and this is the character of most of the movements that are immediately concerned with the maintenance of the vital functions. I might analyze other cases. The act of 'breathing' is a purely reflex action, and goes on when we are perfectly unconscious of exerting any effort, and when our attention is entirely given up to some act or thought; and even when asleep the act of breathing goes on with perfect regularity, and, if it were to stop, of course the stoppage would have a fatal effect upon our lives. But most of these reflex actions are to a certain degree placed under the control of our Will. If it were not for this controlling power of will, I could not be addressing you at this moment. I am able so to regulate my breath as to make it subservient to the act of speech; but that is the case only to a certain point. I could not go on through a long sentence without taking my breath. I am obliged to renew the breath frequently, in order to be able to sustain the circulation and other functions of life. But still I have that degree of control over the act of respiration, that I can regulate this drawing in and expulsion of the breath for the purposes of speech. This may give you an idea of the way in which Mental operations may be independent of the Will, and yet be under its direction. To this we shall presently come.

Now, those reflex actions of the spinal cord, which are immediately and essentially necessary to the maintenance of our lives, take place from the commencement without any training, without any education; they are what we call "instinctive actions;" the tendency to them is part of our nature; it is born with us. But, on the other hand, there are a great many actions which we learn, to which we are trained in the process of bodily education, so to speak, and which, when we have learned them, come to be performed as frequently, regularly, methodically, and unconsciously, as those of which I have spoken. This is the case particularly with the act of walking. You all know with how much difficulty a child is trained to that action. It has to be learned by a long and painful experience, for the child usually gets a good many tumbles in the course of that part of its education; but, when once acquired, it is as natural as the act of breathing, only it is more directly under the control of the will; yet so completely automatic does it become, that we frequently execute a long series of these movements without any consciousness whatever.

You start in the morning, for instance, to go from your home to your place of employment; your mind is occupied by a train of thought, something has happened which has interested you, or you are walking with a friend, and in earnest conversation with him; and your legs carry you on without any consciousness on your part that you are moving them. You stop at a certain point, the point at which you are accustomed to stop, and very often you will be surprised to find that you are there. While your mind has been intent upon something else, either the train of thought which you were following out in your own mind, or upon what your friend has been saying, your legs move on of themselves, just as your heart beats, or as your muscles of breathing continue to act. But this is an acquired habit; this is what we call a "secondarily automatic" action. Now, that phrase is not very difficult when you understand it. By automatic we mean an action taking place of itself. I dare say most of you have seen automata of one kind or another, such as children's toys and more elaborate pieces of mechanism, which, being wound up with a spring, and containing a complicated series of wheels and levers, execute a variety of movements. In each of the Great Exhibitions there have been very curious automata of this kind. We speak, then, of the actions being "automatic," when we mean that they take place of themselves, without any direction on our own parts; such as the act of sucking in the infant, the acts of respiration and swallowing, and others which are entirely involuntary, and are of this purely reflex character.

Now, those are "primarily automatic," that is, originally automatic; we are born with a tendency to execute them; but the actions of the class I am now speaking of are executed by the same portion of the nervous system— the spinal cord—and are "secondarily automatic," that is to say, we have to learn them, but, when once learned, they come very much into the condition of the others, only we have some power of will over them. We start ourselves in the morning by an act of the will; we are determined to go to a particular place; and it may be that we are conscious of these movements over the whole of our walk; but, on the other hand, we may be utterly unconscious of them, and continue to be so until either we have arrived at our journey's end or begin to feel fatigued. Now, when we begin to feel fatigued, we are obliged to maintain the action by an effort of the will; we are no longer unconscious, and we are obliged to struggle against the feeling of fatigue, to exert our muscles in order to continue the action.

Now, having set before you this reflex action of the Spinal Cord, you will ask me perhaps what is the exciting cause of this succession of actions hi walking. I believe it is the contact of the ground with the foot at each movement. We put down the foot, that suggests as it were to the spinal cord the next movement of the leg in advance, and that foot comes down in its turn; and so we follow with this regular rhythmical succession of movements. We next pass to a set of centres somewhat higher, those which form the summit, as it were, of this spinal cord, which are really embedded in the brain, but which do not form a part of that higher organ, which is in fact the organ of the higher part of our mental nature, yet which are commonly included in that which we designate the brain. In fact, the anatomist who only studies the human brain is very liable to be misled in regard to the character of these different parts, by the fact that the higher part—that which we call the Cerebrum—is so immensely developed in Man, in proportion to the rest of the animal creation, that it envelops, as it were, the portion of which I am about to speak, concealing it, and reducing it apparently to the condition of a very subordinate part; and yet that subordinate part is, as I shall show you, the foundation or basis of the higher portion—the Cerebrum itself. The brain of a Fish consists of very little else than a series of these ganglia, these little knots the word "ganglion" means "knot," and the ganglia in many instances, when separated, are little knots, as it were, upon the nerves. The brain of a fish consists of a series of these ganglia, one pair belonging to each principal organ of sense. Thus we have in front the ganglia of smell, then the ganglia of sight, the ganglia of hearing, and ganglia of general sensation. These constitute almost entirely the brain of the fish. There is scarcely any thing in the brain of the fish which answers to the

Cerebrum or higher part of the brain of man. I will give you an idea of the relative development of these parts.

Now, the Cerebrum in most fishes is a mere little film, overlying the sensory tract, but in the higher fish we have it larger; in the reptiles we have it larger still; and in birds we have it still larger; in the lower mammalia it is larger still; and then as we ascend to man this part becomes so large in proportion that my board will not take it in. This Cerebrum, this great mass of the brain, at the bottom of which these Ganglia of Sense are buried, as it were, so overlies and conceals them that their essential functions for a long time remained unknown. Now, in the Cerebrum, the position of the active portion, what we call the ganglionic matter, that which gives activity and power to these nervous centres, is peculiar. In all ganglia this "gray" matter, as it is called, is distinct from the white matter. In ordinary ganglia, this gray matter lies in the interior as a sort of little kernel; but in the Cerebrum it is spread out over the surface, and forms a film or layer. If any of you have the curiosity to see what it is like, you have only to get a sheep's brain and examine it, and you will see this film of a reddish substance covering the surface of the Cerebrum. In the higher animals, and in man, this film is deeply folded upon itself, with the effect of giving it a very much more extended surface, and in this manner the blood-vessels come into relation with it; and it is by the changes which take place between this nervous matter and the blood that all our nervous power is produced. You might liken it roughly to the galvanic battery by which the electric telegraph acts, the white or fibrous portion of the brain and nerves being like the conducting wires of the telegraph. Just as the fibres of the nerves establish a communication between the organs of sensation and the ganglionic centres, and again between the ganglionic centres and the muscles, so do the white fibres, which form a great part of the brain, establish a communication between the gray matter of the convoluted or folded surface of the Cerebrum and the Sensory Ganglia at its base.

Now, I believe that this sensory tract which lies at the base of the skull is the real *Sensorium*, that is, the centre of sensation; that the brain at large, the cerebrum, the great mass of which I have been speaking, is not in itself the centre of sensation; that, in fact, the changes which take place in this gray matter only rise to our consciousness—only call forth our conscious mental activity—when the effect of those changes is transmitted downward to this Sensorium. Now, this

33

Sensorium receives the nerves from the organs of sense. Here, for instance, is the nerve from the organ of smell, here from the eye, and here from the body generally (the nerves of touch), and here the nerves of hearing—every one of these has its own particular function. Now, these Sensory ganglia have in like matter reflex actions. I will give you a very curious illustration of one of these reflex actions: You all know the start we make at a loud sound or a flash of light; the stimulus conveyed through our eyes from the optic nerve to the central ganglion causing it to send through the motor nerves a mandate that calls our muscles into action. Now, this may act sometimes in a very important manner for our protection, or for the protection of some of our delicate organs.

A very eminent chemist a few years ago was making an experiment upon some extremely explosive compound which he had discovered. He had a small quantity of this compound in a bottle, and was holding it up to the light, looking at it intently; and whether it was a shake of the bottle or the warmth of his hand, I do not know, but it exploded in his hand, the bottle was shivered into a million of minute fragments, and those fragments were driven in every direction. His first impression was, that they had penetrated his eyes, but to his intense relief he found presently that they had only penetrated the outside of his eyelids. You may conceive how infinitesimally short the interval was between the explosion of the bottle and the particles reaching his eyes; and yet in that interval the impression had been made upon his sight, the mandate of the reflex action, so to speak, had gone forth, the muscles of his eyelids had been called into action, and he had closed his eyelids before the particles reached them, and in this manner his eyes were saved. You see what a wonderful proof this is of the way in which the automatic action of our nervous apparatus enters into the sustenance of our lives, and the protection of our most important organs from injury.

Now I have to speak of the way in which this Automatic action of the Sensory nerves, and of the motor nerves which answer to them, grows up, as it were, in ourselves. We will take this illustration: Certain things are originally instinctive, the tendency to them is born with us; but in a very large number of things we educate ourselves, or we are educated. Take, for instance, the guidance of the class of movements I was speaking of just now—our movements of locomotion. We find that when we set off in the morning with the intention of going to our place of employment, not only do our legs move without our consciousness, if we are attending to something entirely different, but we guide ourselves in our walk through the streets; we do not run up against anybody we meet; we do not strike ourselves against the lamp-posts; and we take the appropriate turns which are habitual to us. It has often happened to myself, and I dare say it has happened to every one of you, that you have intended, to go somewhere else—that when you started you intended instead of going in the direct line to which you were daily accustomed, to go a little out of your way to perform some little commission; but you have got into a train of thought and forgotten yourself, and you find that you are half-way along your accustomed track before you become aware of it. Now, there, you see, is the same automatic action of these sensory ganglia—we see, we hear—for instance, we hear the rumbling of the carriages, and we avoid them without thinking of it—our muscles act in respondence to these sights and sounds—and yet all this is done without our intentional direction—they do it for us. Here again, then, we have the "secondarily automatic" action of this power, that of a higher nervous apparatus which has grown, so to speak, to the mode in which it is habitually exercised. Now, that is a most important consideration. It has grown to the mode in which it is habitually exercised; and that principle, as we shall see, we shall carry into the higher class of Mental operations.

But there is one particular kind of this action of the Sensory nerves to which I would direct your attention, because it leads us to another very important principle. You are all, I suppose, acquainted with the action of the stereoscope; though you may not all know that its peculiar action, the perception of solidity it conveys to us, depends upon the combination of two dissimilar pictures—the two dissimilar pictures which we should receive by our two eyes of an object if it were actually placed before us. If I hold up this jug, for instance, before my eyes, straight before the centre of my face, my two eyes receive pictures which are really dissimilar. If I made two drawings of the jug, first as I see it with one eye, and then with the other, I should represent this object differently. For instance, as seen with the right eye, I see no space between the handle and the body of the jug; as I see it with the left eye, I see a space there. If I were to make two drawings of that jug as I now see it with my two eyes, and put them into a stereoscope, they would bring out, even if only in outline, the conception of the solid figure of that jug in a way that no single drawing could do. Now, that conception is the result of our early-acquired habit of combining with that which we see that which we *feel*. That habit is acquired during the first twelve or eighteen months of infancy. When your little children are lying in their cradles and are handling a solid object, a block of wood, or a simple toy, and are holding it at a distance from their eyes, bringing it to their mouth and then carrying it to arm's length, they are going through a most important part of their education—that part of their education which consists in the harmonization of the mental impressions derived from sight and those derived from the touch; and it is by that harmonization that we get that conception of solidity or projection which, when we have once acquired it, we receive from the combination of these two dissimilar pictures alone, or even, in the case of objects familiar to us, without two dissimilar pictures at all—the sight of the object suggesting to us the conception of its solidity and of its projection.

Now, this is a thing so familiar to you, that few of you have probably ever thought of reasoning it out; and in fact it has only been by the occurrence of cases in which persons have grown to adult age without having acquired this power, from having been born blind and having only received sight by a surgical operation at a comparatively late period, when they could describe things as they saw them— I say it is only by such cases that we have come to know how completely dissimilar and separate these two classes of impressions really are, and how important is this process of early infantile education of which I have spoken. A case occurred a few years ago in London where a friend of my own performed an operation upon a young woman who had been born blind, and, though an attempt had been made in early years to cure her, that attempt had failed. She was able just to distinguish large objects, the general shadow, as it were, of large objects without any distinct perception of form, and to distinguish light from darkness. She could work well with her needle by the touch, and could use her scissors and bodkin and other implements by the training of her hand, so to speak, alone. Well, my friend happened to see her, and he examined her eyes, and told her that he thought he could get her sight restored; at any rate, it was worth a trial. The operation succeeded; and, being a man of intelligence and quite aware of the interest of such a case, he carefully studied and observed it; and he completely confirmed all that had been previously laid down by the experience of similar cases. There was one little incident

which will give you an idea of the education
which is required for what you would suppose
is a thing perfectly simple and obvious. She
could not distinguish by sight the things that
she was perfectly familiar with by the touch, at
least when they were first presented to her
eyes. She could not recognize even a pair of
scissors. Now, you would have supposed that
a pair of scissors, of all things in the world,
having been continually used by her, and their
form having become perfectly familiar to her
hands, would have been most readily
recognized by her sight; and yet she did not
know what they were; she had not an idea
until she was told, and then she laughed, as
she said, at her own stupidity. No stupidity at
all; she had never learned it, and it was one of
those things which she could not know
without learning. One of the earliest cases of
this kind was related by the celebrated

Cheselden, a surgeon of the early part of last century. Cheselden relates how a youth just in this condition had been accustomed to play with a cat and a dog; but for some time after he attained his sight he never could tell which was which, and used to be continually making mistakes. One day being rather ashamed of himself for having called the cat the dog, he took up the cat in his arms and looked at her very attentively for some time, stroking her all the while; and in this way he associated the impression derived from the sight of the cat with the impression derived from the touch, and made himself master (so to speak) of the whole idea of the animal. He then put the cat down, saying, "Now, puss, I shall know you another time."

Now, the reason why I have specially directed your attention to this is because it leads to one of the most important principles

that I desire to expound to you this evening—
what I call in Mental Physiology the doctrine
of *resultants*. All of you who have studied
mechanics know very well what a "resultant"
means. You know that when a body is acted
on by two forces at the same time, one force
carrying it in this direction, and another force
in that direction, we want to know in what
direction it will go, and how far it will go. To
arrive at this we simply complete what is
called the parallelogram of forces. In fact, it is
just as if a body were acted on at two different
times, by a force driving it in one direction,
and then by a force driving it in the other
direction.

We draw two lines parallel to this, and we
draw a diagonal—that diagonal is what is
called the resultant; that is, it expresses the
direction, and it expresses the distance—the
length of the motion which that body will go

when acted upon by these two forces. Now, I use this term as a very convenient one to express this—that when we have once got the conception that is derived from the harmonization of these two distinct sets of impressions on our nerves of sense, we do not fall back on the original impressions, but we fall back on the resultant, so to speak. The thing has been done for us; it is settled for us; we have got the resultant; and the combination giving that resultant is that which governs the impression made upon our minds by all similar and future operations of the same kind. We do not need to go over the processes of judgment by which the two sets of impressions are combined in every individual case; but we fall back, as it were, upon the resultant. Now, what is the case in the harmonization of the two classes of impressions of sight and touch, I believe to be true of the far more complicated operations of the mind of which the higher portion of the brain, the Cerebrum, is the instrument. Now, this Cerebrum we regard as furnishing, so to speak, the mechanism of our thoughts. I do not say that the Cerebrum is that which does the whole work of thinking, but it furnishes the mechanism of our thought. It is not the steam-engine that does the work; the steam-engine is the mere mechanism; the work is done, as my friend Prof. Roscoe would tell you, by the heat supplied; and if we go back to the source of that heat, we find it originally in the heat and light of the sun that made the trees grow by which the coal was produced, in which the heat of the sun is stored up, as it were, and which we are now using, I am afraid, in rather wasteful profusion.

The steam-engine furnishes the mechanism; the work is done by the force. Now, in the same manner the brain serves as the mechanism of our thought; and it is only in that sense that I speak of the work of the brain. But there can be no question at all that it works of itself, as it were,—that it has an automatic power, just in the same manner as the sensory centres and the spinal cord have automatic power of their own. And that a very large part of our mental activity consists of this automatic action of the brain, according to the mode in which we have trained it to action, I think there can be no doubt whatever. And the illustration with which I started in this lecture gives you, I believe, a very good example of it. However, there are other examples which are in some respects still better illustrations of the automatic work that is done by the brain, in the state which is sometimes called Second Consciousness or Somnambulism—to which some persons are peculiarly subject. I heard only a few weeks ago of an extremely remarkable example of a young man who had overworked himself in studying for an examination, and who had two distinct lives, as it were, in each of which his mind worked quite separately and distinct from the other. One of these states, however— the ordinary one—is under the control of the will to a much greater extent than the other; while the secondary state is purely, I suppose, automatic. There are a great many instances on record on very curious mental work, so to speak, done in this automatic condition a state of active dreaming, in fact. For instance, Dr. Abercrombie mentions, in his very useful work on "The Intellectual Powers," an example of a lawyer who had been excessively perplexed about a very complicated question. An opinion was required from him, but the question was one of such difficulty that he felt very uncertain how his opinion should be given. The opinion had to be given on a certain day, and he awoke in the morning of that day with a feeling of great distress. He said to his wife, "I had a dream, and the whole thing in that dream has been clear before my mind, and I would give anything to recover that train of thought." His wife said to him, "Go and look on your table." She had seen him get up in the night and go to his table and sit down and write. He went to his table, and found there the very opinion which he had been most earnestly endeavoring to recover, lying in his own handwriting. There was no doubt about it whatever, and this opinion he at once saw was the very thing which he had been anxious to be able to give. A case was put on record of a very similar kind only a few years ago by a gentleman well known in London, the Rev. John De Liefde, a Dutch clergyman. This gentleman mentioned it on the authority of a fellow-student who had been at the college at which he studied in early life. He had been attending a class in mathematics, and the professor said to his class one day: "A question of great difficulty has been referred to me by a banker, a very complicated question of accounts, which they have not themselves been able to bring to a satisfactory issue, and they have asked my assistance. I have been trying, and I cannot resolve it. I have covered whole sheets of paper with calculations, and have not been able to make it out. Will you try?" He gave it as a sort of problem to his class, and said he should be extremely obliged to anyone who would bring him the solution by a certain day. This gentleman tried it over and over again; he covered many slates with figures, but could not succeed in resolving it. He was a little put on his mettle, and very much desired to attain the solution; but he went to bed on the night before the solution, if attained, was to be given in, without having succeeded. In the morning, when he went to his desk, he found the whole problem worked out in his own hand. He was perfectly satisfied that it was his own hand; and this was a very curious part of it—that the result was correctly obtained by a process very much shorter than any he had tried. He had covered three or four sheets of paper in his attempts, and this was all worked out upon one page, and correctly worked, as the result proved. He inquired of his "hospita," as she was called—I believe our English equivalent is bedmaker, the woman who attended to his rooms—and she said she was certain that no one had entered his room during the night. It was perfectly clear that this had been worked out by himself.

Now, there are many cases of this kind, in which the mind has obviously worked more clearly and more successfully in this automatic condition, when left entirely to itself, than when we have been cudgelling our brains, so to speak, to get the solution. I have paid a good deal of attention to this subject, in this way: I have taken every opportunity that occurred to me of asking inventors and artists—creators in various departments of art—musicians, poets, and painters, what their experience has been in regard to difficulties which they have felt, and which they have after a time overcome. And the experience has been almost always the same, that they have set the result which they have wished to obtain strongly before their minds, just as we do when we try to recollect something we have forgotten; they think of everything that can lead to it; but, if they do not succeed, they put it by for a time, and give their minds to something else, and endeavor to obtain as complete a repose or refreshment of the mind upon some other occupation as they can; and they find that either after sleep, or after some period of recreation by a variety of employment, just what they want comes into their heads. A very curious example of this was mentioned to me a few years ago by Mr. Wenham, a gentleman who has devoted a great deal of time and attention to the improvement of the microscope, and who is the inventor of that form of binocular microscope (by which we look with two eyes and obtain a stereoscopic picture) which is in general use in this country.

The original binocular microscope was made upon a plan which would suggest itself to any optician. I shall not attempt to describe it to you, but it involved the use of three prisms, giving a number of reflections; and every one of these reflections was attended with a certain loss of light and a certain liability to error. And, besides that, the instrument could only be used as a binocular microscope. Now, Mr. Wenham thought it might be possible to construct an instrument which would work with only one prism, and that this prism could be withdrawn, and then we could use the microscope for purposes to which the binocular microscope could not be applied. He thought of this a great deal, but he could not think of the form of prism which would do what was required. He was going into business as an engineer, and he put his microscopic studies aside for more than a fortnight, attending only to his other work, and thinking nothing of his microscope. One evening after his day's work was done, and while he was reading a stupid novel, as he assured me, and was thinking nothing whatever of his microscope, the form of the prism that should do this work flashed into his mind. He fetched his mathematical instruments, drew a diagram of it, worked out the angles which would be required, and the next morning he made his prism, and found it answered perfectly well; and upon that invention nearly all the binocular microscopes made in this country have since been constructed.

I could tell you a number of anecdotes of this kind which would show you how very important is this automatic working of our minds—this work which goes on without any more control or direction of the Will, than when we are walking and engaged in a train of thought which makes us unconscious of the movements of our legs. And I believe that in all these instances—such as those I have named, and a long series of others—the result is owing to the mind being left to itself without the disturbance of any emotion. It was the worry which the bank-manager had been going through, that really prevented the mind from working with the steadiness and evenness that produced the result. So in the case of the lawyer; so in the case of the mathematician; they were all worrying themselves, and did not let their minds have fair play. You have heard, I dare say, and those of you who are horsemen may have had experience, that it is a very good thing sometimes, if you lose your way on horseback, to drop the reins on the horse's back and let him find his way home. You have been guiding the horse into one path and into another, and following this and that path, and you find that it does not lead you in the right direction; just let the horse go by himself, and he will find his way better than you can. In the same manner, I believe that our minds, under the circumstances I have mentioned, really do the work better than our wills can direct. The will gives the impulse in the first instance, just as when you start on your walk; and not only this, but the will keeps before the mind all the thoughts which it can immediately lay hold of, or which association suggests, that bear upon the subject. But then these thoughts do not conduct immediately to an issue, they require to work themselves out; and I believe that they work themselves out very often a great deal better by being left to themselves. But then we must recollect that such results as these are only produced in the mind which has been trained and disciplined; and that training and discipline are the result of the control of the Will over the mental processes, just as in the early part of the lecture I spoke to you of the act of speech as made possible by the control which the will has over the muscles of breathing. We cannot stop these movements— we must breathe—but we can regulate them, and modify them, and intensify them, or we can check them for a moment, in accordance with the necessities of speech.

Well, so it is, I think, with regard to the action of our will upon our mental processes. I believe that this control, this discipline of the will, should be learned very early; and I will give to the mothers among you, especially, one hint in regard to a most valuable mode of training it even in early childhood. I learned this, I may say, from a nurse whom I was fortunate enough to have, and whose training of my own sons in early childhood I regard as one of the most valuable parts of their education. She was a sensible country girl, who could not have told her reasons, but whose instincts guided her in the right direction. I studied her mode of dealing: with the children, and learned from that the principle. Now, the principle is this: A child falls down and hurts itself. (I take the most common of nursery incidents. You know that Sir Robert Peel used to say that there were three ways of looking at this question; and there are three modes of dealing with this commonest of nursery incidents.) One nurse will scold the child for crying. The child feels the injustice of this; it feels the hurt, and it feels the injustice of being scolded. I believe that is the most pernicious of all the modes of dealing with it. Another coddles the child, takes it up and rubs its head, and says, "O naughty chair, for hurting my dear child!" I remember learning that one of the royal children fell against a table in the queen's presence, and the nurse said, "O naughty table," when the queen very sensibly said: "I will not have that expression used; it was not the table that was naughty; it was the child's fault that he fell against the table." I believe that this method is extremely injurious; the result of it being that it fixes the child's attention upon its hurt, and causes it to attain that habit of self-consciousness which is in after-life found to have most pernicious effects. Now, what does the sensible and judicious nurse do? She distracts the child's attention, holding it up to the window to look at the pretty horses, or gets it a toy to look at. This excites the child's attention, and the child forgets its hurt, and in a few moments is itself again, unless the hurt has been severe. When I speak of coddling, I mean about a trifling hurt such as is forgotten in a few moments; a severe injury is a different matter. But I believe that the coddling is only next in its evil results (when followed out as a system) to the evil effects of the system of scolding; the distraction of the attention is the object to be aimed at. Well, after a time the child comes to be able to distract its own attention. It feels that it can withdraw its own mind from the sense of its pain, and can give its mind to some other object, to a picture-book or to some toy, or whatever the child feels an interest in; and that is the great secret of self-government in later life. We should not say, "I won't think of this"—some temptation, for instance; *that* simply fixes the attention upon the very thought that we wish to escape from; but the true method is, "I will think of something else;" *that*, I believe, is the great secret of self-government, the knowledge of which is laid in the earliest periods of nursery-life.

Now, just direct your attention to this diagram, as a sort of summary of the whole:

[DIAGRAM]

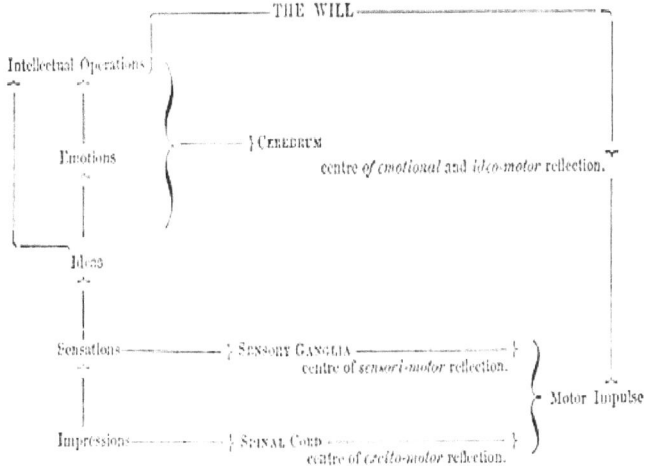

You see I put at the top the Will. The will dominates everything else. I do not pretend to explain it, but I simply say, as Archbishop Manning said, in applying my own language to this case, that our common-sense teaches us that we have a will, that we have the power of self-government and self-direction, and that we have the power of regulating and dominating all these lower tendencies to a certain extent, not to an unlimited extent. We cannot prevent those thoughts and feelings rising in our minds that we know to be undesirable; but we can escape from them, we can repress them; but, as I said, the effort to escape from them is much more effectual than the effort to repress them, excepting when they arise with great power, and then we have immediately, as it were, to crush them out; but when they tend to return over and over again, the real mode of subduing them is to determine to give our attention to something else. It is by this exercise of the will, therefore, in training and disciplining the mind, that it acquires that method by which it will work of itself. The mathematician could never have worked out that difficult problem, nor the lawyer have given his opinion, nor the artist have developed those conceptions of beauty which he endeavors to shape either in music, or poetry, or painting, but for the training and disciplining which his mind has undergone. The most wonderfully creative of all musicians, Mozart, whose music flowed from him with a spontaneousness that no musician, I think, has ever equalled—Mozart went through, in early life, a most elaborate course of study, imposed upon him, in the first instance, by his father, and afterward maintained by himself. When his contemporaries remarked how easily his compositions flowed from him, he replied, "I gained the power by nothing but hard work." Mozart had the most extraordinary combination of this intuitive musical power, with a knowledge derived from patient and careful study, that probably any man ever attained. Now, in the same manner we have persons of extraordinary natural gifts, and see these gifts frequently running to waste, as it were, because they have not received this culture and discipline. And it is this discipline which gives us the power of performing, unconsciously to ourselves, these elaborate mental operations; because I hold that a very large part of our mental life thus goes on, not only automatically, but even below the sphere of our consciousness. And you may easily understand this if you refer to the diagram which I drew just now on the black-board. You saw that the Cerebrum, the part that does the work, what is called the convoluted surface of the brain, lies just immediately under the skull-cap; that it is connected with the sensorium at the base of the brain by a series of fibres which are merely, I believe, conducting fibres. Now, I think that it is just as possible that the Cerebrum should work by itself when the sensorium is otherwise engaged or in a state of unconsciousness, as that impressions should be made on the eye of which we are unconscious. A person may be sleeping profoundly, and you may go and raise the lid and bring a candle near, and you will see the pupil contract; and yet that individual shall see nothing, for he is in a state of perfect unconsciousness. His eye sees it, so to speak, but his mind does not; and you know that his eye sees it by the contraction of the pupil, which is a reflex action; but his mind does not see it, because the sensorium is in a state of inaction. In the same manner during sleep the Cerebrum may be awake and working, and yet the Sensorium shall be asleep, and we may know nothing of what the cerebrum is doing except by the results. And it is in this manner, I believe, that, having been once set going, and the cerebrum having been shaped, so to speak,

67

in accordance with our ordinary processes of mental activity, having grown to the kind of work we are accustomed to set it to execute, the cerebrum can go on and do its work for itself. The work of invention, I am certain, is so mainly produced, from concurrent testimony I have received from a great number of inventors, or what the old English called "makers"—what the Greeks called poets, because the word poet means a maker. Every inventor must have a certain amount of imagination, which may be exercised in mechanical contrivance or in the creations of art; these are *inventions*—they are made, they are produced, we don't know how; the conception comes into the mind we cannot tell whence; but these inventions are the result of the original capacity for that particular kind of work, trained and disciplined by the culture we have gone through. It is not given to every one of us to be an inventor. We may love art thoroughly, and yet we may never be able to evolve it for ourselves. So in regard to humor. For instance, there are some men who throw out flashes of wit and humor in their conversation, who cannot help it—it flows from them spontaneously. There are other men who enjoy this amazingly, whose nature it is to relish such expressions keenly, but who cannot make them themselves. The power of invention is something quite distinct from the intellectual capacity or the emotional capacity for enjoying and appreciating; but although we may not have these powers of invention, we can all train and discipline our minds to utilize that which we do possess to its utmost extent. And here is the conclusion to which I would lead you in regard to Common-Sense. We fall back upon this, that common-sense is, so to speak, the *general resultant of the whole previous action of our minds.* We submit to common-sense any questions—such questions as I shall have to bring before you in my next lecture; and the judgment of that common-sense is the judgment elaborated as it were by the whole of our mental life. It is just according as our mental life has been good and true and pure, that the value of this acquired and this higher common-sense is reached. We may in proportion I believe to our honesty in the search for truth—in proportion as we discard all selfish considerations and look merely at this grand image of truth, so to speak, set before us, with the purpose of steadily pursuing our way toward it—in proportion as we discard all low and sensual feelings in our love of beauty, and especially in proportion to the earnestness of the desire by which our minds are pervaded always to keep the right before us in all our judgments—

so I believe will our minds be cleared in their perception of what are merely prudential considerations. It has on several occasions occurred to me to form a decision as to some important change either in my own life, or in the life of members of my family, which involved a great many of what we are accustomed to cal *pros* and *cons*—that is, there was a great deal to be said on both sides. I heard the expression once used by a naturalist, with regard to difficulties in classification—"It is very easy to deal with the white and the black; but the difficulty is to deal with the gray." And so it is in life. It is perfectly easy to deal with the white and the black there are things which are clearly right, and things which are clearly wrong; there are things which are clearly prudent, and things which are clearly imprudent; but a great many cases arise in which even right and wrong may seem balanced, or the motives may be so balanced that it is difficult to say what is right; and again, there are cases in which it is difficult to say what is prudent; and I believe, in these cases where we are not hurried and pressed for a decision, the best plan is to do exactly that which I spoke of in the earlier part of the lecture—to set before us as much as possible everything that is to be said on both sides. Let us consider this well; let us go to our friends; let us ask what they think about it. They will suggest considerations which may not occur to ourselves. It has happened to me within the last three or four months to have to make a very important decision of this kind for myself; and I took this method—I heard everything that was to be said on both sides, I considered it well, and then I determined to put it aside as completely as possible for a month, or longer, if time should be given, and then to take it up again, and simply just to see how my mind gravitated—how the balance then turned. And I assure you that I believe that in those who have disciplined their minds in the manner I have mentioned, that act of "Unconscious Cerebration," for so I call it, this unconscious operation of the brain in balancing for itself all these considerations, in putting all in order, so to speak, in working out the result—I believe that process is far more likely to lead us to good and true results than any continual discussion and argumentation, in which one thing is pressed with undue force, and then that leads us to bring up something on the other side, so that we are just driven into antagonism, so to speak, by the undue pressure of the force which we think is being exerted. I believe that to hear everything that is to be said, and then not to ruminate upon it too long, not to be continually thinking about it, but to put it aside entirely from our minds as far as we possibly can, is the very best mode of arriving at a correct conclusion. And this conclusion will be the *resultant* of the whole previous training and discipline of our minds. If that training and discipline has all been in the direction of the true and the good, I believe that we are more likely to obtain a valuable result from such a process than from any conscious discussion of it in our minds, anything like continually bringing it up and thinking of it, and going over the whole subject again in our thought. The unconscious settling down, as it were, of all these respective motives will, I think, incline the mind ultimately to that which is the just and true decision.

There is just one other point I could mention in connection with this subject: the manner in which the *conscious* direction and discipline of the mind will tend to remove those *unconscious* prejudices that we all have more or less from education, from the circumstances in which we were brought up; and from which it is excessively difficult for us to free ourselves entirely. I have known a great many instances, in public and in private life, in which the most right-minded men have every now and then shown the trammelling, as it were, of their early education and early associations, and were not able to think clearly upon the subject in consequence of this. These early prejudices and associations dinar around us and influence the thoughts and feelings of the honestest men in the world unconsciously; and it is sometimes surprising, to those who do not know the force of these early associations, to see how differently matters which are to them perfectly plain and obvious are viewed by men whom we feel we must respect and esteem. Now, I believe that it is the earnest habit of looking at a subject from first principles, and, as I have said over and over again, looking honestly and steadily at the true and the right, which gives the mind that direction that ultimately overcomes the force of these early prejudices and these early associations, and brings us into that condition which approaches the nearest of anything that I think we have the opportunity of witnessing in our earthly life, to that *direct insight*, which many of us believe will be the condition of our minds in that future state in which they are released from all the trammels of our corporeal existence.

www.ingramcontent.com/pod-product-compliance
Lightning Source LLC
Chambersburg PA
CBHW071823200526
45169CB00018B/850